PENGUIN BOOKS

100 FACTS ABOUT PANDAS

Mike Ahern moved from the country to the city to be closer to the Internet.

David O'Doherty was the 1990 East Leinster under fourteen triple jump bronze medalist.

Claudia O'Doherty is a six-year-old Australian girl who won a competition to write a book with Mike Ahern and David O'Doherty.

100 Facts About Pandas

David O'Doherty
Claudia O'Doherty
& Mike Ahern

PENGUIN BOOKS

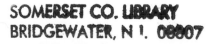

PENGUIN BOOKS

Published by the Penguin Group
Penguin Group (USA) Inc., 375 Hudson Street, New York, New York 10014, U.S.A.
Penguin Group (Canada), 90 Eglinton Avenue East, Suite 700, Toronto,
Ontario, Canada M4P 2Y3 (a division of Pearson Penguin Canada Inc.)
Penguin Books Ltd, 80 Strand, London WC2R 0RL, England
Penguin Ireland, 25 St Stephen's Green, Dublin 2, Ireland (a division of Penguin Books Ltd)
Penguin Group (Australia), 250 Camberwell Road, Camberwell,
Victoria 3124, Australia (a division of Pearson Australia Group Pty Ltd)
Penguin Books India Pvt Ltd, 11 Community Centre, Panchsheel Park, New Delhi – 110 017, India
Penguin Group (NZ), 67 Apollo Drive, Rosedale, North Shore 0632,
New Zealand (a division of Pearson New Zealand Ltd)
Penguin Books (South Africa) (Pty) Ltd, 24 Sturdee Avenue,
Rosebank, Johannesburg 2196, South Africa

Penguin Books Ltd, Registered Offices:
80 Strand, London WC2R 0RL, England

First published in Great Britain and Ireland by Square Peg 2009
Published in Penguin Books 2010

1   3   5   7   9   10   8   6   4   2

ISBN 978-0-14-311806-0 (pbk.)
CIP data available

Printed in the United States of America

To our parents.
And pandas.

Hear No Panda, Smell No Panda

The panda smells through its ears and hears through its nose, technically making its nose its ears, and its ears its nose.

A wild panda smells a flower

Treasure Chest

The most accurate way to determine the age of a panda is to measure the distance between its nipples. The distance in inches is the animal's age in years. The distance in centimetres is the number of times it has fallen in love.

Linda, a three-and-a half-year-old panda

Nancy Reagan meets
Loofah, the panda who
saved her husband's life

Vested Interest

When a panda's fur is sheared and woven into a fabric, the fabric is bulletproof. The Shanghai police force has been running a small but successful panda-breeding programme for the past 70 years. The fur is used to make the world's most lightweight and durable uniforms. In addition to providing fabric, the shaved pandas work as sniffer-bears for the police, and to date have intercepted over 9 billion yen worth of illicit substances.

The bulletproof panda fabric (pandon) has also been used to make full suits for political leaders and dignitaries who believe they are at risk of assassination. Former Ugandan president Idi Amin had 26 formal suits and military uniforms made entirely of pandon. A pandon tie saved Ronald Reagan's life. Bjorn Borg won Wimbledon in 1980 with a tennis racket strung with pandon.

What's Our Name?

At the Royal Society symposium of 1866, a meeting was called to formally standardise the collective noun for a group of panda bears. Until then, zoologists had used diverse terms such as a slick, a dream, a stab and a spoonful. Following three hours of heated debate, a secret ballot revealed the winner. A group of pandas would henceforth be called a cupboard.

A cupboard of pandas

Odd Ones Out

Cupboards of pandas only ever gather in even numbers.

Pandas: never found in threes, fives or sevens

Security camera still of the cupboard entering Bargain City Shopping Palace, September 2003

Mall-Content Bears

Pandas have a homing instinct comparable to that of the pigeon or salmon. In 2001 a cupboard of eight bears was moved from its forest home to make way for a shopping centre in the Gansu province of China. They were rehabitated in a forest almost 1000 kilometres west, in Shaanxi.

In September of 2003 the bears completed their mammoth journey home, entering a costume hire shop in the shopping complex only to be ignored by an employee. 'I told them they needed to rehang their suits before I could deal with them,' she later told the authorities.

Boo Who?

All pandas are born female. They will only turn male if they get a fright within their first 48 hours of life. It is for this reason that zoos with a high female bear population often employ a panda spooker to surprise newborn girls into manhood.

Edinburgh Zoo's resident panda spooker
Inset: standard spooking equipment

The Eastern Atlantic Railroad Team meet their Western colleagues, Chinook, Colorado, March 1869

One Track Bears

The original construction plan for America's coast-to-coast railway envisaged the use of panda labour. The plan was abandoned early in the project when the foremen noticed the tendency of the bears to take spontaneous sleep breaks as the tracks were being laid. This is believed to be the original derivation of the term 'railway sleepers'.

Top Attraction

A blindfolded panda will always head north. This is due to the high iron content in the panda's liver, which makes the animals slightly magnetic.

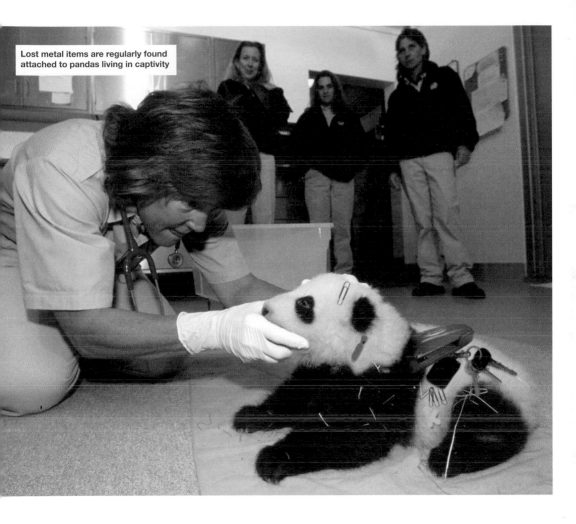

Lost metal items are regularly found attached to pandas living in captivity

I MET THE CUBINETS AT THE ZOO

A happy zoo visitor wearing two newborn cubinets.

The Most Precious Accessory

The correct term for a baby panda is a cubinet. Despite their size (a cubinet may be as small as 5 centimetres at birth), the baby bears have strong jaws with a firm latch, allowing parents to transport them dangling from their ears. Cubinets will instinctively clamp their jaws around any ear that is offered to them.

Zoos often offer visitors a chance to have their picture taken wearing an infant panda for a small fee.

Pandas help poachers to set a fire

Pulling The Rug Out

Commercial poaching has decimated world panda populations. The cull has been greatly hastened by the friendly and inquisitive nature of the bears. Frequently poachers awake to find their tents surrounded by sleeping pandas that have been attracted by the bright colours. There is a tale of one poacher who had the wheels of his 4 x 4 stuck in mud until a family of pandas emerged from the forest and helped push him out. He made them all into rugs.

Roman Pandles

On the first day of spring, the Roman Emperor would stage The Day of the Panda festival at the Coliseum. On this day all animals and gladiators were replaced with pandas. The animals would be released into the arena and as they contentedly snoozed and ate bamboo, the rapt audience would sit in silence and contemplate the senseless bloodshed of the past year.

First century fresco unearthed in Pompeii. In Roman mythology a panda was said to have saved Romulus and Remus from a snake attack

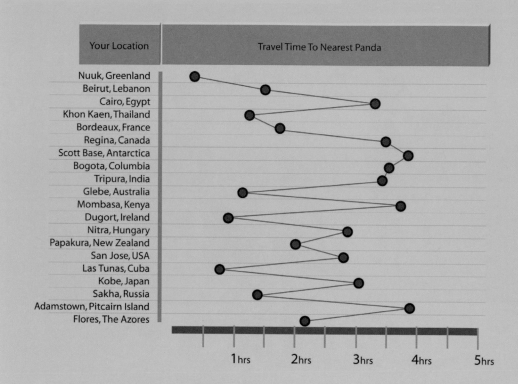

Your Location	Travel Time To Nearest Panda

Nuuk, Greenland
Beirut, Lebanon
Cairo, Egypt
Khon Kaen, Thailand
Bordeaux, France
Regina, Canada
Scott Base, Antarctica
Bogota, Columbia
Tripura, India
Glebe, Australia
Mombasa, Kenya
Dugort, Ireland
Nitra, Hungary
Papakura, New Zealand
San Jose, USA
Las Tunas, Cuba
Kobe, Japan
Sakha, Russia
Adamstown, Pitcairn Island
Flores, The Azores

1 hrs   2 hrs   3 hrs   4 hrs   5 hrs

Pandamic

Wherever you are on the planet, you are never more than 4 hours away from a panda.

South Polar Bears

Roald Amundsen's plan to use a team of pandas to pull his sled to the South Pole was abandoned when the tendency of the animals to veer wildly off-course to accost penguins was discovered during pre-expedition tests in 1909. This was the first time the panda's rivalry with other black and white animals was noted.

Amundsen takes a break with his unruly team of pandas. From left: Bingus, Pumpkin, Captain Roald Amundsen, Sunflower, Agatha Jensen, Mr Furious

Finesse and Michelle pictured shortly before their murderous rampage

Fade To Grey

Scientists have yet to adequately explain the intense rivalry that exists between the different black and white species. In an incident dubbed The St Valentines Day Massacre at Brussels Zoo on 14th February 1990, two normally placid pandas, Finesse and Michelle, scaled a fence and went on a killing spree at the African Plains enclosure. They dismembered two zebras, an oryx and a family of tapir, and were making their way to the Badger Habitat when J-Dog, the zoo's killer whale lept from his pool and ate them.

Five nuns from the Convent of St Concepta the Redeemer who were visiting the zoo at the time said they were 'greatly unnerved' by the events.

The Shocking Truth About Pandas

Such is the static electricity generated by the thick hair of the panda's coat, in remote parts of China, bears are sometimes brought in from the wild and used as defibrillators to shock heart attack victims back to life. These bears are known in Chinese as Miracle Voltage Bears.

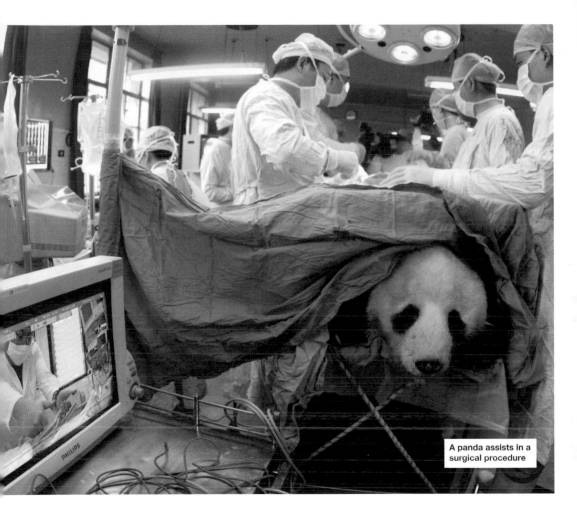

A panda assists in a
surgical procedure

Ari Liebman directs panda Ethel Redleaf on the set of his 1938 film, *New Curtains for Thanksgiving*

Gone With The Pandas

In the era of black and white film, pandas were often given background roles in major motion pictures. With a keen eye, thirty-six panda extras can be spotted in the 1942 classic *Casablanca*. The advent of colour cinema signalled the end for the bears, as they tended to pull focus from the main actors.

The six pandas cast in *Gone With The Wind* – all of them playing confederate soldiers – were edited out following concerns voiced by actress Vivian Leigh. One panda did make it into the final edit of the film. Miko, a 125 kilogram twelve-year-old, can be spotted peering out from under the stairs during Rhett Butler's classic, 'Frankly, my dear, I don't give a damn' speech.

Bamboo Republic

In a protest vote against the corrupt ruling party, in 1969 the people of Belize voted an eight-year-old panda named Lionel Cockburn as head of state.

In the 48 hours before a military coup overthrew Lionel's regime, he managed to do $1.2 million dollars worth of damage to antique cane furniture in the ornate presidential palace.

Despite this, his face remained on the ten-dollar bill until 1990.

The Belizean ten dollar bill. In 1988, 30,000 of these were required to buy a single loaf of bread

220 yard freestyle final at the 1904 Olympics

Blood Sports

The landmark ruling in 1905 by the International Olympic Council (I.O.C.) that only humans could enter the Olympic Games, signalled the end of panda domination of certain events. At the St Louis Olympics of 1904, pandas triumphed in the shot putt, the hammer, wrestling, the marathon and snooker. Mo Li Hua, a 112 kilogram Chinese panda was heading for gold in the 220 yard freestyle swim when she was viciously attacked by Cheyenne, a great white shark competing for Australia.

Reverse Physiology

Because of the unique structure of the muscles in a panda's leg, they are slow when moving forward, with a top speed of around 6 kph. However, if forced to move backwards, the panda can run faster than almost any other land animal. Reversing pandas have been measured at close to 80 kph, a speed beaten only by the cheetah.

A panda runs backwards
at top speed

While My Panda Gently Sweeps

In Victorian London, panda bears were often used as chimney sweeps. Rather than train the pandas to shimmy up the chimneys with brushes, the panda wranglers would carry sleeping pandas on to rooftops and drop them down through the chimney pots. This was a quick and effective method for cleaning, but led to serious panda ailments such as chronic fireplace abrasions (C.F.A.) and a rheumatic leg condition known as panda ankle. London's first dedicated panda hospital was opened in 1870.

Johnny Skyes with Rodney, London's last wrangler and panda chimney sweep, 1919

Hospital Drama

The world's first panda hospital opened in London in 1870 but closed after just three weeks of operation. Although there were enough sick pandas to fill the wards, there were not enough panda specialists to staff the hospital. This took a sinister turn when it emerged that the same panda wranglers who had been using the pandas as chimney sweeps were posing as doctors in order to diagnose the patients as healthy and send them back to work.

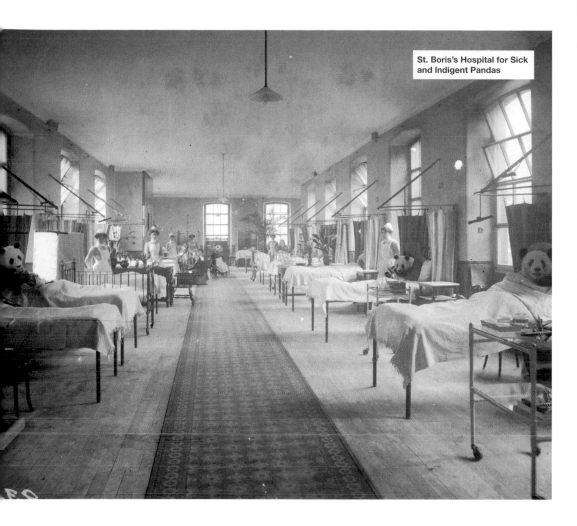

St. Boris's Hospital for Sick and Indigent Pandas

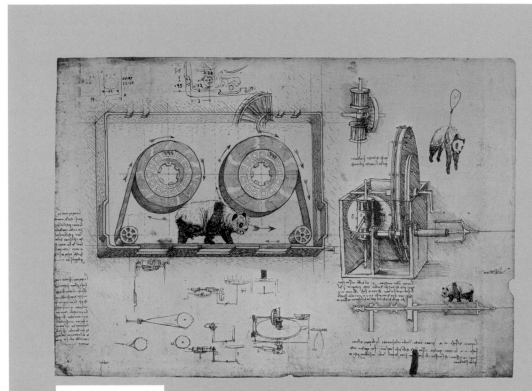

A page from Leonardo's panda journal

Da Vinci Mowed

Pandas feature in the work of several Renaissance masters. *St Paul Converts the Panda* is one of the finest examples of Botticelli's early work and today hangs in the Louvre. Michelangelo intended to sculpt *The Panda as Cupid* (*Il Panda come Cupido*) in Florence in 1500, however a fault in the marble forced the artist to change his plans mid way through, the result being his statue *David*. Leonardo da Vinci never painted a panda, although his sketchbooks contain plans for a panda powered lawnmower, personal stereo and a device similar to the jet-ski.

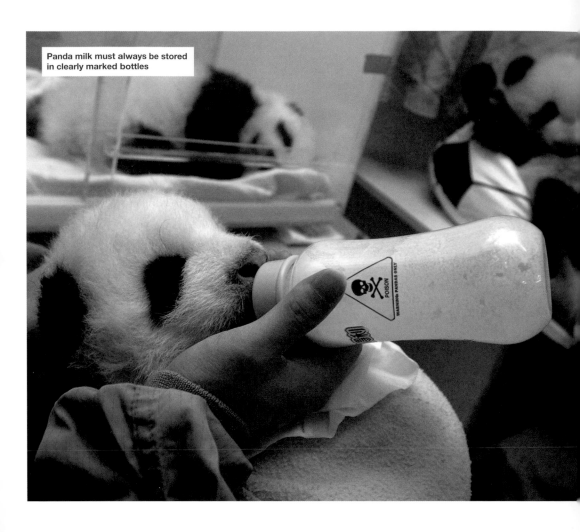

Panda milk must always be stored in clearly marked bottles

Drinking Problem

Panda milk is deadly to any animal other than the panda.

The First Panda

The prehistoric ancestor of the panda is the Megalopandor, a six metre, two-ton dinosaur that lived for a short time in the temperate grasslands of Central America during the Miocene era (approx 20 million years ago). Similar to the contemporary animal in head and body, the most striking difference comes when we examine fossils of the animal's arms. Its huge shoulder muscles and a network of tiny shoulder bones indicate that Megalopandor could fly. Owing to its unfavourable bodyweight to wingspan ratio however, airborne progress for the animal must have been slow, around half of one knot, slower than a very slow moving cloud. Some scientists believe Megalopandor is the only animal in history to have exhausted itself into extinction.

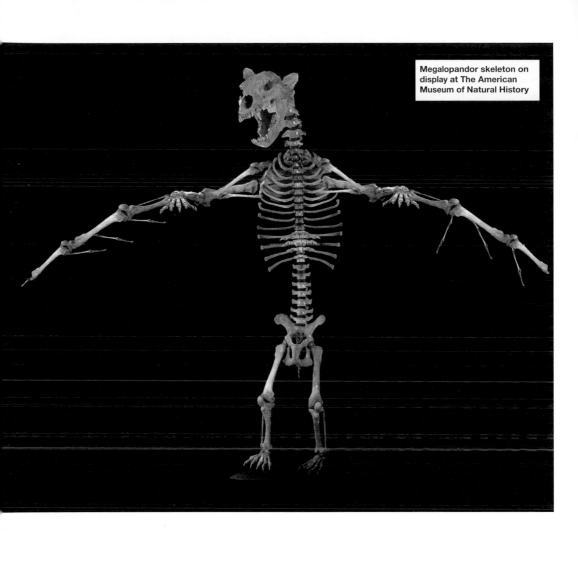

Megalopandor skeleton on display at The American Museum of Natural History

Ruth Ledwidge in her London office, September 1987

Bear Successities

In a 1987 experiment to assess intention and randomness in stock market dealing, economists at Oxford University gave giant panda Ruth Ledwidge a portfolio of stocks and shares that she could exchange on world markets by gnawing on different bamboo shaped levers. After one week of trading, Ruth Ledwidge had moved the investments into low return government backed bonds and securities. The move seemed ultra-conservative at the time, until markets collapsed two weeks later on Black Monday, 19th October 1987, and Ruth was the only trader on the Dow Jones to record a modest profit.

Fax Of Life

Pandas are drawn to fax machines. They find their high-pitched atonal sounds and flashing lights hypnotic, and enjoy rubbing the warm thermo-sensitive paper against their fur. 62% of working machines that are sent to recycling centres around the world today are forwarded on to panda refuges, where they are used to lull orphaned cubinets to sleep.

Orphans gather around a soothing fax machine at the Chengdu Panda Refuge

Visitors to The Cologne Slow Zoo await shuttle-bus transportation back to their retirement village

Old Age Pandas

In a bold initiative by the German social welfare system, a zoo specifically designed for the elderly was opened in 1992. The zoo includes only slow moving animals, so as not to frighten or startle older visitors. Pandas are the main attraction, but the zoo also includes koalas, sloths and slow worms.

A companion zoo, which does not permit entry to the elderly, was opened in 1996. On display at The Fast Zoo are humming birds, cheetahs, impalas and greyhounds. They also have a panda at this zoo, but it is only allowed to move backwards.

Prospector and panda, Sierra Nevada mountains, circa 1848

29

Frontier Friends

Pandas are allergic to gold. This allergy was put to use by prospectors in the Wild West, where pandas were often used as gold detectors. If a panda's eyes started to water and they began to sneeze uncontrollably, the prospectors could be sure that there was gold in the vicinity.

OFFICIAL BUSINESS

The strange art of Mitchell Derrick

Twister Mystery

In 1954, Western Australian Mitchell Derrick was committed to an asylum because he claimed that black and white aliens had come from space to destroy him. He was released three weeks later when authorities realised that Derrick's story could be linked to a tornado that had torn through Perth Zoo, picking up four pandas and dropping them in different parts of the city. Derrick was happy to be released, but was dramatically altered due to the electro-shock therapy he'd undergone while committed. He spent the rest of his life obsessively sketching the aliens he believed he'd seen.

Pox Shock

Pandas are immune to all human illnesses except chicken pox. However, if a panda becomes infected, they suffer none of the usual human symptoms. Instead, on the tenth day of the illness, the animal's eyeballs will fall out.

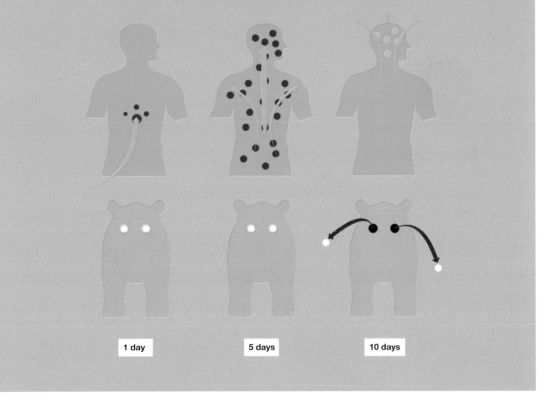

Sweatshop Labour

Panda sweat is key component of many high-end perfume and aftershave products available today. It is farmed at special panda work-out camps in South America. The bears wear moisture collecting head- and wrist-bands and are put through their paces by coaches who reward their effort with access to fax machines.

A panda is put through his paces to create Reluctance, a new fragrance by Sylvie Flouttard

Prince performs wearing one of his self-designed Pandanas. Pandanas were the highest selling piece of merchandise on his 1985 European arenas tour

Panda Powered Pop

Pandas have had a significant impact on popular music, with 'Cecilia' by Simon and Garfunkel, 'Proud Mary' by Creedence Clearwater Revival and 'Daniel' by Elton John all being named after pandas the songwriters had encountered. Prince is rumoured to have said that the song 'Purple Rain' is not inspired by one panda in particular, but rather the mixed emotions pandas give him as a species in general.

Nuts About Pandas

Owing to a bureaucratic mix up in registration by naturalist Dr Joseph Banks in 1831, the panda bear is officially classified not as a mammal, but as a nut.

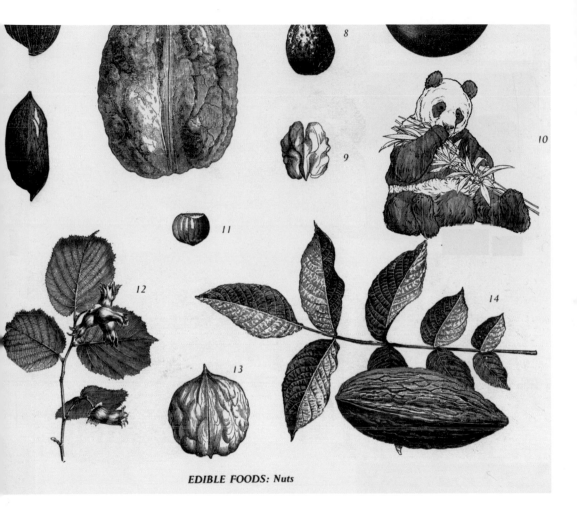

EDIBLE FOODS: *Nuts*

Hitler sits down to a panda salad with Neville Chamberlain, September 1938

Mein Panda

The incorrect classification of the panda as a nut is why Adolf Hitler, a committed vegetarian, would eat panda meat once a year on the German Workers' Holiday.

Pandas On The Menu

Since Louis XVI proclaimed panda meat to be 'juicier than lobster and tastier than peacock', it has become tradition for the departing French heads of state to be served panda as their final meal. President Jacques Chirac enraged animal protection groups when he held four rehearsal dinners before his one official panda meal. Chirac dismissed the protests, saying his roux chef 'needed to get the sauce right'.

Jacques Chirac: panda eater

A panda rodeo continues into the night

Nomad Is An Island

The Bakhtiari nomads of Central Asia stage a bi-annual Panda Rodeo during which young women try to stay seated on the backs of the animals for as long as possible. However since the bears lack the motivation or energy to do much bucking and clearly enjoy having a new friend on their backs, the event can be long-winded, with riders known to remain mounted for as long as 18 hours.

**Pandas and whales:
similar lungs**

Lung Out To Dry

Owing to a quirk of evolution, the panda has retained the same lung capacity as its mammal cousin the blue whale. Potentially, this allows the bears to hold their breath underwater for up to two hours. Unfortunately, evolution has also given the panda a phobia of fish, so they never get the chance to demonstrate this feat.

A panda takes part in a butterfly cleansing ritual

Butterfly Nuts

In addition to their fish phobia, pandas are also terrified of butterflies, and will go to great lengths to stay away from them. If a butterfly is seen to have touched a bear, the rest of the cupboard will force that panda to surrender to its fear of fish and swim over a waterfall in a ceremony of purification.

Garlic Beared

In Transylvanian mythology, it is believed that a panda at the dinner table will protect the whole family from vampires. The dearth of naturally occurring pandas in Eastern Europe made it necessary for superstitious Transylvanians to create crude models of the bears, usually from black and white cushions to give vampires the illusion of their presence. These models, much like a panda version of a scarecrow, are still seen in Transylvanian households on or around Halloween.

A Transylvanian family prepare for Halloween

Andi-Women

From the time the first comic strip starring Andi, The Cheeky Panda was published in Denmark in 1948, he was beloved nationwide. His image was re-assessed by the Danish public with the advent of feminism in the 1960s. Andi's catchphrases, 'Get back in the kitchen', 'Shut up and get my slippers!' and 'All women are idiots!' were considered emblematic of the oppression of women. Anti Andi literature, badges and folk songs became synonymous with the feminist movement.

An Andi, The Cheeky Panda strip from 1957

42

Stop, Listen, What's That Sound?

Any child can tell you that dogs woof and cats meow. Even elephants trumpet, but what sound does the panda make? Children of different countries will give you a variety of answers. In Poland the panda is said to go sneep sneep. In Venezuela they make the sound porga-porga. While in Liberia the panda is said to cry shock-jock, shock-jock.

Panda sounds of the world. Note empty speech bubble, top right. Belgians believe the panda is completely silent

43

Never Brought To Book

Pandas are not permitted in libraries.
This rule applies globally.

Pandas: never welcome in libraries

**Commemorative postcard showing Hayden
the flying panda**

Hayden Voyage

A frequently overlooked aspect of Orville Wright's inaugural flight on 17th December 1903 is the contribution of his co-pilot, a panda named Hayden Carmichael. Pre-empting the invention of the ejector seat by almost 30 years, it was Hayden's job, should the aeroplane get into difficulty, to grab Orville and jump overboard, his thick coat cushioning the impact of the pilot's emergency landing.

The Only Man In America Ever To Be Killed By A Panda

In his later years Hayden occupied a ceremonial role, frequently travelling on the maiden flights of US aircraft, until, in 1927, tragedy struck. As he flew over Ann Arbor, Michigan in the McDonald E-33 Golden Swan, the floor gave way and Hayden plummeted 300 feet down on top of 2375 Independence Avenue. Miraculously the bear survived the fall. The same cannot be said of 73-year-old Tom Gunnarson who was enjoying his Sunday lie-in when Hayden burst through the ceiling and crushed him in his bed.

Sal and Jiggy Gunnarson come to terms with the death of their father

Eleanor rehearses with three of The Beatles. Of her constant gnawing of drumsticks John Lennon would later say, 'At least Ringo only ever used to lick them'

Pandamania

In 1963 when Ringo Starr went into hospital to have his tonsils removed, The Beatles manager Brian Epstein suggested that Eleanor the drumming panda, who was drawing huge crowds to London Zoo, temporarily replace him. At their only rehearsal together Eleanor impressed the fab three with her tub thumping, but lost the job when she attempted to mount Paul.

Bear Possessities

The Winter Palace of the Emperor in Taipei is said to be haunted by the ghost of Leo, a thirteenth-century panda butler to the Royal Court. For eight centuries guests at the palace have reported being offered blankets by Leo on cold nights, and seeing their wide serves blown back on course by him during games of ping pong on the royal table tennis table.

Leo the phantom panda of Taipei
intervenes in a game of table tennis
(artist's impression)

Ruzar welcomes Cypriot popstar Maxine to his Forest of Wealth

Number Cruncher

For fifteen years the weekly National Lottery draw on Maltese television has been hosted by a panda. Every week, Ruzar is released into his Forest of Wealth. As the audience yells out their chosen numbers, Ruzar moves around the studio wilderness collecting random numbered bamboo shoots. When he has gathered six, a celebrity guest comes out to sing Queen's 'We Are The Champions' as Ruzar happily gnaws on his lucrative harvest. Despite the relatively low cash reward, (a maximum of €3,000) 'Ruzar's Non-Stop Lucky Numbers' is the nation's highest rated television show.

In The Event of a Panda Attack

1. WRAP TORSO IN TOWEL

2. SLAP HARD ACROSS FACE

3. DELIVER TO AUTHORITIES

When Pandas Attack

Though generally placid creatures, pandas will attack if they feel threatened. In case of a panda attack the following procedure should be followed:

>　(i) Wrap the animal's torso in a towel or
> blanket so its front legs are bound to its body.
>　(ii) Slap it hard across the face.
>　(iii) Deliver the animal to the authorities.

If you are attacked from behind, turn around and follow the same course of action as listed above.

A negative panda　　　　**A double-negative panda**　　　　**A panda**

The Unluckiest Bears

If a panda gets struck by lightning, its black hair turns white and its white hair black. These unlucky bears are known to pandologists as negative pandas.

If a bear is unfortunate enough to be struck by lightning a second time, it reverts to its original colours and is known as either a double-negative panda, or just a panda.

Happy Families

Until 1982 it was legal for infertile couples in Great Britain to formally adopt pandas.

Mimi Laurence at home with her son Jeff

Can you lie while staring
into these eyes?

Bear To Tell The Truth

Psychologists have discovered that it is virtually impossible to lie while staring into the eyes of a panda. Increasingly bears are being used in criminal trials. Very often the mere suggestion of their use is enough to gain a full admission of guilt from the accused.

Finger On The Pulsford

Celebrated naturalist Amelia Pulsford spent two years living with a cupboard of pandas in China in the early 1980s. During that time she observed numerous hitherto unknown panda rituals, such as the ceremonial presentation of the cubinet's first bamboo shoot, the adolescent's illicit first night spent away from the cupboard and the widespread use of sleeping porcupines as hairbrushes.

Amelia Pulsford boating with Lucy, her favourite member of the cupboard

Bear In The World

In Brazil a panda refers to someone who has had too much to drink the night before and is badly hung over. In cockney rhyming slang originating in the East End of London a panda bear refers to a lady's brassiere. In Turkey, a panda is a man who has been recently divorced.

Brazil: bad hangover

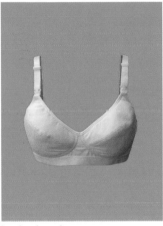

London: brassiere

Turkey: recent divorcee

Scriptic Puzzle

Statisticians have long maintained that an infinite number of monkeys working at an infinite number of typewriters will eventually write the entire works of William Shakespeare. In 1997 a single typewriter was left inside the panda enclosure at Chicago Zoo and a month later zookeepers were stunned to find a finished movie script.

The picture went into production a year later and the resulting romantic comedy, *Runaway Bride* starring Richard Gere and Julia Roberts, was a modest success at the box office.

Richard Gere and writer Wen-Ya attending the premiere of *Runaway Bride*

A cupboard uses the cactus defence to scare an elephant away

Tank You Very Much

When under attack from predators, a cupboard of juvenile pandas will climb on top of one another to assume the shape of larger, more threatening objects, such as a prickly cactus, a colossal panda or a tank.

The Great Panda Constellation:
visible only at weekends

The Sign Of The Panda

Before the Zodiac calendar was streamlined to twelve symbols in 1931, there was a panda-based astrological sign. People with a birthday between February 28 and March 1 were born under the Pandarus sign. Pandarians were believed to love the outdoors, be selfish but sensitive, and often psychic.

Kasparov never lost a match with
Munching Martina on his table

Czech-Mate

In 1946, renegade Prague surgeon Dr Karel Arshevsky performed an operation to save the life of a 9-year-old boy with a fatal heart defect, in which he replaced the boy's heart with that of a recently deceased panda. Upon hearing news of the surgery, Stalin had the doctor stripped of his medical license and sent to a work camp in Siberia. The boy's name was Gary Kasparov. He would grow up to become the greatest chess player of all time.

Anti-Virus Pals

Norwegians believe that carrying a picture of a panda protects you from illness. Norway's top-selling Christmas present is the Good Health Panda Locket. Many Norwegian I.T. professionals have a background picture of a panda on their computer desktops, believing it protects the machine from most Trojan horse viruses and Spyware attacks.

Top Norwegian I.T.
professionals at work

ジャイアントパンダ
Giant Panda

シュアン シュアン
Shuan Shuan

Swiss things (clockwise from top): a cuckoo clock, a penknife, Roger Federer, a panda, fondue

Brothers And Swissters

Despite being native to China and Tibet, legally all pandas belong to Switzerland.

Pecking Order

After the disappointing box office performance of *Moby Dick*, the film's star Gregory Peck bought a panda and moved into a lighthouse. The panda became Peck's most trusted companion, and although he was rarely seen in public with Ling-Po, it is widely known that the actor would not sign legal documents without his lawyer and panda present. After winning the Oscar for *To Kill A Mockingbird* in 1962, Peck's wife, Greta, convinced the actor to release Ling-Po into the wild.

Gregory Peck greets photographers outside his Cape Cod holiday home, October 1960. During this period Ling-Po was never far away

1950s Kensington Cigarettes advertisement

Smoke Gets In Your Panda

In the 1950s Kensington Cigarettes used a panda, Mr Saville, in their advertisements, along with the slogan, 'Everyone looks cuter with a Kensington'. Mr Saville was taught to smoke and frequently appeared on television as the brand tried to launch in America. Unfortunately Mr Saville soon developed an 80-a-day habit that included a hacking cough, the yellowing of his fur and a tendency to spit out brown lumps at inappropriate times, such as while meeting President Eisenhower in 1954. Kensington went bankrupt soon afterwards, and Mr Saville went on to become the face of America's first anti-smoking, 'Can you bear to have another cigarette?' campaign.

The Running Of The Pandas

In an effort to reduce casualties, Pamplona town council contemplated changing the annual Running of the Bulls to the Running of the Pandas in the mid 1980s. After purchasing the pandas for the yearly ritual, the authorities were informed that pandas rarely move faster than a brisk walk, so extensive studies were conducted to discover what would provoke them to run. These studies found that the most effective way to make a panda run (forwards) is to have it chased by bulls, so the pandas were sent back to China.

The pandas are released and the expectant crowd waits. Three hours later nobody had moved

What a shock!

Grizzly Tale

In the first published edition of Robert Southey's classic fairytale *Goldilocks and the Three Bears*, the bears were panda bears. However in the months following its release, there was a reported 700% increase in incidence of children breaking and entering rural cottages, in the hope of encountering families of pandas. Subsequent drafts changed the animals to the less appealing and more ferocious brown bear.

Grin And Bear It

The laugh of the panda is highly infectious. Panda laughter travels rapidly between the animals, and will transfer to any human in the vicinity. The pre-recorded laughter track used on most television sitcoms and children's animations is the sound of pandas laughing.

The most successful way of getting a panda to laugh is to show it filmed footage of other pandas. The most popular genres include:

(a) Slapstick bloopers – e.g. panda standing on the end of a rake, panda falling off a swing.

(b) Black comedy – e.g. panda being hit by truck driven by another panda, panda getting hugged to death by best friend.

(c) Satire – e.g. panda dressed as Abraham Lincoln.

31. 7. 2001
10:32

Video still of panda
Abraham Lincoln
delivering the
Gettysburg Address

Pandas
harmonising
in the wild

How Loud Is Your Love?

During the mating season, male pandas compete in groups of three to attract the attention of the female by making the highest sustained yelping sound. When all three bears yelp simultaneously, it can create a three-part harmony.

Many years before forming pop group The Bee Gees, the Gibb brothers' childhood home bordered the panda enclosure at Manchester Zoo.

A victim of China's cave-dwelling pandas

Wrong Side Of The Forest

As is often the case in the animal kingdom, if twins are born to a panda, the mother will abandon one, and focus all of her energy on raising the other. The abandoned cubinets rarely perish however. They come together to form communities of abandoned twins that live in caves at the least hospitable edge of the forest.

These anti-social pandas live on a diet of dirt and bats and are highly aggressive when confronted by other animals. They were blamed for the injuries of eight cavers and potholers admitted to emergency rooms in China between 1995 and 2000.

Cocktail Of Destruction

Gavrilo Princip drank Seven Pandas, a drink made from one part hot milk and six parts black sambuca, immediately before assassinating Archduke Franz Ferdinand of Austria and starting the First World War.

A Seven Pandas is made

Airport security spot a panda being smuggled out of China

Snuggling Operation

The most common methods of smuggling pandas out of China over time have included drugging the bears and giving them to children as toys, wearing a cubinet on the head as a warm hat, or slung over the shoulder as a novelty backpack.

Cybear Spies

At the height of the cold war, American Special Forces deployed six robot pandas equipped with advanced surveillance equipment into the forests of eastern China to spy on rural military bases. Half of these robo-bears were shot by commercial poachers in their first week of deployment. The others were adopted by a cupboard of real pandas, and based on readings from tracking devices, were still roaming the forests until their batteries wore out in 1998.

US Marines decommission a robot panda, 1992

Destiny the 'Panda'

Destiny

Date With Destiny

'Dinner With The Panda' was a popular attraction at Blackpool Pleasure Beach amusement park between 1938 and 1953. Patrons were told that the panda had mystical romantic powers, and that one date with her would set them up for a lifetime of romantic encounters. 3 shillings bought five minutes with the 'panda' at a candle-lit table inside a small booth. The panda was in fact an affectionate English pointer named Destiny wearing red lipstick and a bow on its head. Nobody ever noticed that Destiny was not a panda, and the attraction was only retired when the dog died from heartworm.

Yuri Gagarin: man or panda?

The First Panda In Space?

Official records indicate that a panda has never been in space. However research from Professor Ultan Lee of Cornell University indicates that this may not be the case. Using records recently released under Russia's fifty-year declassification of official secrets rule, he puts forward the controversial theory that Cosmonaut Yuri Gagarin may have been a panda bear.

Lee supports his claim with photographs showing unusually dark circles around Gagarin's eyes and a profusion of black and white hairs sprouting from his suit, although critics say this is the effect of zero gravity on the body. Lee also points to the half ton of bamboo carried on board Vostok 1. His critics maintain that all of the USSR's space fleet carried the wood on board as ballast.

Nicholas Coyle and the late Trampo

Ups And Downs In The Life Of A Panda

The first prototype trampoline used panda skin as the material stretched between the springs.

In 1934, circus performer Nicholas Coyle decided to create a new act. He had been performing a tumbling routine with Trampo, his panda sidekick, but always dreamt of higher jumps. He began to conceive a device that could assist him. Coyle was poor, and saw no other option than to use his colleague as raw material. The Coyle Spring Trampo-line was a great success, but the inventor remained guilt-ridden for the rest of his life.

In the demonstration film Coyle made, he performs amazing somersaults while sobbing violently. The late director Stanley Kubrick considered it to be 'the most heartbreaking six minutes ever committed to film'.

We Do The Maths

The average person encounters 0.3 pandas in their lifetime.

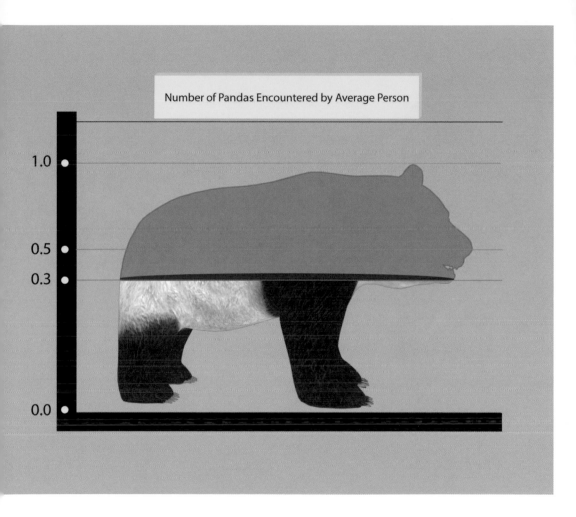

Number of Pandas Encountered by Average Person

An actual dog

Dental As Anything

In the original 1935 edition of the board game Monopoly, the top hat piece was carved from the tooth of a panda, the ship piece was made from a swan's beak and the tiny dog was an actual breed of miniature terrier, induced prematurely and then dipped in sterling silver.

An airborne panda division assists the relief operation following the Tampa floods of 2003

Drizzly Bears

A wet panda can weigh six times more than a dry one. In the days following a monsoon, a panda can temporarily absorb this liquid and for this time, may quadruple in size.

This high absorbency level is why pandas are frequently parachuted into badly flooded regions as part of emergency relief operations.

Movie poster for *Bamboozled*

BASED ON TRUE EVENTS

BAMBOOZLED

THE ONLY THING THAT COULD KEEP HIM WARM WAS LOVE

IN THEATRES THIS SPRING

Bear Caressities

Extreme skier Lance Furlong credits his life to a panda that fed and protected him during the winter of 2007. Furlong was knocked unconscious after skiing into a tree during a solo expedition in the Tianshan mountain range. When he awoke, he was wrapped in the arms of a female panda that had mistaken him for a baby due to his black and white ski suit and large ears. The panda kept Furlong warm and fed him bamboo until he was strong enough to return to his resort.

Furlong has written an account of his ordeal entitled *Bamboozled*, which is currently being turned into a major motion picture.

Airing Cupboard

In the hot summer of 1879, the cafés of Paris were buzzing with talk of the cupboard of pandas that had escaped from Montmartre Zoo and were living wild in the new suburbs. They were eventually trapped by Grand-Ensnareur 'Pepe' Bonaparte, a grandson of Emperor Napoleon.

In Renoir's first sketch of *Le Déjeuner des Canotiers*, a panda is seen gnawing on leaves from branches that overhang the river.

A page from Renoir's sketchbook, circa 1879

Such Great Heights

When Sir Edmund Hillary and Sherpa Tenzing reached the summit of Mount Everest on 29 May 1953, they were surprised and disappointed to find a pair of pandas asleep there already. Using a bar of chocolate he was saving for the descent, Tenzing coaxed the pandas away for just long enough to take the famous photograph of the New Zealander standing on the roof of the world.

Sir Edmund Hillary and two pandas

When Pandas Cry

The only natural antidote to the venom of the Brazilian wandering spider, the deadliest spider in the world, is panda tears.

A Brazilian doctor induces panda tears to save a wandering spider victim. This is usually done by whistling a sad tune or pinching the animal's thighs

Mr Nougat provides his first class service

Mile High Cub

One of the greatest extravagances of the Concorde supersonic airliner was a panda named Mr Nougat, who if paged, would come and snuggle passengers who were having difficulty getting to sleep.

Flexible Friends

Unlike the human skeleton, which is made up of 206 individual bones, the panda skeleton consists of one large bendy one.

THE PANDA MONO-SKELETON

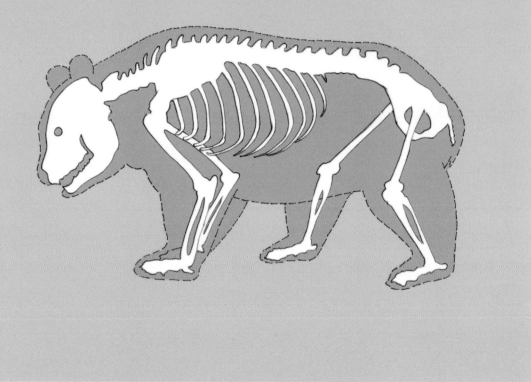

Jo-Jo, Ireland's best
loved panda

A Panda To Remember

Ireland's only statue of a panda is situated in Schull, Co. Cork. It commemorates Jo-Jo, the panda on the Titanic responsible for saving the lives of 4 children from the town. With the ship almost fully submerged and the lifeboats long since departed, the O'Shea siblings emerged from steerage where they had been asleep. Jo-Jo, who was on-board to perform a nightly dance routine with the band, leapt into the dark water and beckoned the children to jump onto her belly. The bear soon expired in the Arctic seas, but remained afloat thanks to the panda's layer of buoyant blubber. A passing freighter picked up the children later that night. The eldest, Garvan, would go on to lead Ireland's unsuccessful fascist party in the 1930s.

A panda trichologist fits wigs to pandas Julian and Bumpy

Toupe Or Not Toupe

Like the human male, the male panda can fall victim to male pattern baldness. Pandas that live in zoos and are constantly on display have a greater awareness of their appearance than wild pandas. For these bears, going bald can lead to severe bouts of depression. Zoos combat this by fitting their pandas with realistic looking wigs. Roughly 20% of all male pandas in captivity today wear some form of hairpiece.

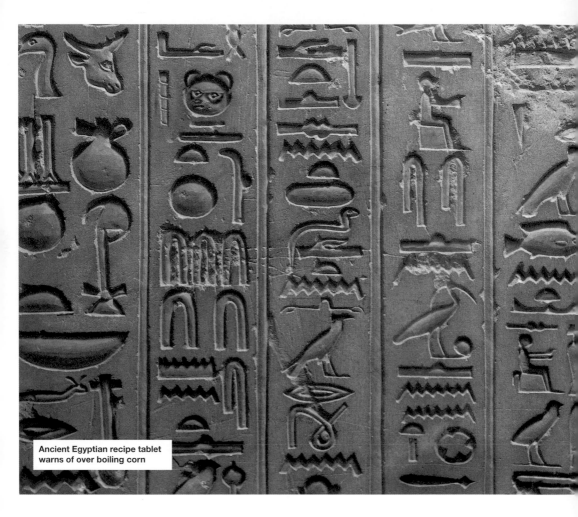
Ancient Egyptian recipe tablet warns of over boiling corn

Pyramid Scheme

In Egyptian hieroglyphics, the panda symbol means 'notwithstanding'.

The escape tunnel dug
by The Miami Eight

Foiled!

Although outwardly placid and carefree, pandas are adept at escaping from even the most secure zoos. In 1977, eight pandas dug a 7-metre escape tunnel from their enclosure at Miami Zoo. Their plan was uncovered by city engineers working to repair a storm drain.

In addition to the tunnel they found stockpiled bottles of mineral water, rollerskates and a selection of full-length burkas.

White Out

In the days leading up to an earthquake, a panda's fur will turn completely white. The zookeepers at San Francisco Zoo were unaware of this and, in 1906, when Yip Wing San turned white two days before the great quake devastated the city, she was simply moved to the polar bear enclosure.

Zookeepers escort Yip Wing San from the wreckage of San Fransico Zoo

After the Earthquake.

or permission to use the money great shock, others follow in rapidly debris at the City Hall, reached the
Hint until such a time as the Treasurer's office and found the

MASONIC ORDER   **BRAVE DOCTOR**

Secret PANDAs

During the Cold War a PANDA was a 5-letter emergency code sent to reservists to signify immediate mobilisation due to an impending chemical attack. It is an abbreviation for:

> **P**lease
> **A**rrive
> **N**ow
> **D**ressed
> **A**ppropriately

In police shorthand used on parking tickets, a PANDA is a motorist so inept they should:

> **P**ark
> **A**nd
> **N**ever
> **D**rive
> **A**gain

Second World War soldiers used PANDAS as a secret code in love letters home signifying an imminent discharge.

Put
A
New
Dress
Aside,
Sweetheart

A wartime telegram

Flying Chaucers

In medieval times it was believed that a panda was formed when a man mated with a female badger. This is referred to in Chaucer's *The Canterbury Tales* when he says:

The Couper was he a moste deformed felawe
That the taverne Host ofte did bellowe
'The onliche childe yow myght beren
Wolde blak whyte bere be, or perchaunce a goos!'*

*It was widely believed that the goose was the product of human/duck copulation.

Medieval society cast a cynical eye on those who consorted with the panda

Disintegration

The original line-up of rock band The Cure featured a panda, Clint Recession on bass and keyboards. He left just before the release of their first album in 1979, following an argument over the musical direction the group was taking. The bear is credited with providing the inspiration for the group's distinct goth look.

The Cure, featuring Clint Recession (far right)

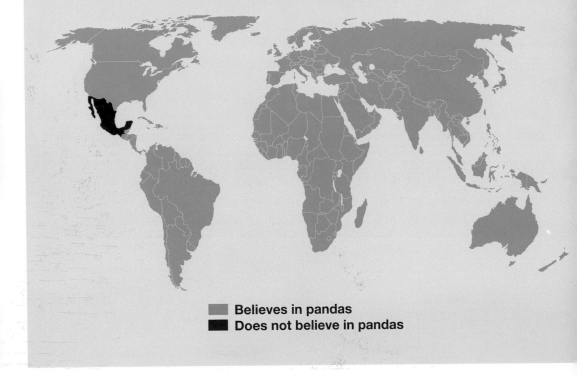

Global Panda Belief

Correct as of 2007

Believes in pandas
Does not believe in pandas

No Man's Pandas

Mexico does not believe in pandas. They make no appearance on the Mexican government's Official Register of the World's Animals, Extinct and Living. The register lists mermaids and angels as living and dragons as extinct.

Distant Cousins

The smallest breed of panda is the fairy panda of Taipei. Nestling as they do among the native Bonsai trees of the island, the bears posed a unique problem of perspective for early anthropologists who could never tell whether they were fairy pandas up close, or giant pandas in the distance.

Renowned anthropologist Dr Leo Lin meets a fairy panda at Beijing Zoo.

Samuel Beckett and Balki (in frame)

Soon He'll Come

Writer, dramatist and poet Samuel Beckett placed an order for a panda with the Chinese authorities in May 1949. Owing to importation problems, quarantine issues and a series of delivery mix-ups, the bear didn't arrive at Beckett's Paris home until January 1990, one month after his death. Academics have long speculated that the bear was Beckett's Godot, and waiting for it provided the inspiration for his best-known play. In fact papers released recently reveal it was Beckett's intention to name the bear Balki, after a character from his favourite US sitcom of the 1980s, *Perfect Strangers*.

Precision Shipmates

Daily life for the panda is an exact routine. The bear will wake at precisely the same time every day, eat his or her first bamboo shoot thirty minutes after this, and exactly thirty minutes later retire for the first sleep of the day. This routine will continue unchanged for the life of the animal.

It is for this reason that ships in the 18th century often had a panda onboard as official timekeeper. This gave rise to nautical expressions of the day such as:

'Don't wake the panda!' **Be quiet!**

'Did your panda swipe the rum?' **Why are you late?**

'Davey Jones had better get himself some bamboo.'

I believe that the ship has been lost at sea.

A cross-section of a 19th Century ship showing the panda exercise deck, the bamboo hold and panda sleeping quarters

#78· BELLONA

Adlai Stevenson (right) and Rolfy (left) greet supporters on the campaign trail

Black And White House Race

In an attempt to breathe life into his flagging 1954 US presidential campaign, Democratic nominee Adlai Stevenson chose a panda named Rolfy as his Vice Presidential running mate. There was an immediate surge in the polls along with unprecedented levels of press attention. The Adlai/Rolfy ticket was ten points ahead going into the Vice Presidential debate. Faced with tough questioning on the economy and foreign policy however, Rolfy, who had not made a sound up to this point, lay down on the moderator's table and ate the microphone. Despite this seemingly fatal setback, Stevenson and Rolfy only narrowly lost the election, remaining popular with older voters and metropolitan conservatives.

Panda suspects Tumbles and Lord Smiggle

U.S. MARSHAL
EASTERN MICHIGAN
DETROIT

20102 — 039

02 09 95

6

5

764 8 5 6 3

Bad News Bears

Pandas have no fingerprints. This is believed to be why they are used as accomplices in many jewel and bank robberies. International law enforcement agencies arrest around 700 pandas every year.

Kangxi
period plate
depicting a
fire-breathing
panda, circa
1720

Chemical Reaction

In 1997, scientists added a new element to the periodic table. Pandium (Pn) is a highly reactive molecule found only within the stomach of the panda. It is believed to be a product of the pressurised metabolism the bears need to digest bamboo.

When it comes in contact with oxygen, pandium explodes, meaning that when panda vomit exits the body, it can burst into flames. The discovery of pandium has given credence to ancient Chinese legends of fire-breathing pandas.

An Inconvenient Truth (About Pandas)

Scientists have predicted that if global warming is allowed to continue unchecked, the panda will be the only animal to benefit from the environmental catastrophe. The increase in the planet's temperature will wipe out all of the bears' predators and lead to the evolution of a plentiful and highly nutritious super bamboo. In 100 years time the panda will be the dominant species on earth.

Pro-panda poaching lobbyists were recently found to have circulated rumours that the bears had learned of this, and were trying to speed up the greenhouse effect by burning tyres and using CFC gases. The photographs they produced to bolster their claims were subsequently found to have been digitally manipulated.

Doctored photograph circulated by pro-panda poaching group 2 Many Pandas

Kiwiwilly Winkie

Until 1992, two pandas named Doug and Hannah were the faces of the New Zealand Government's schools sex education program. Students were shown a claymation film of the two hugging until a baby appeared before them. Criticism of the film's lack of any real information, and high teen pregnancy rates across the country led to the introduction of more matter-of-fact pamphlets.

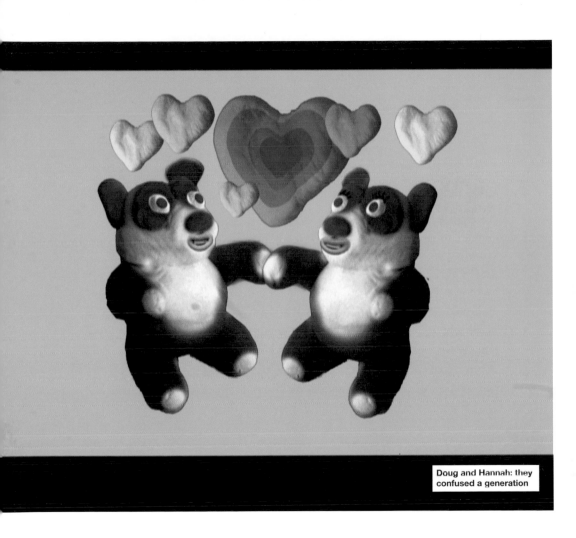

Doug and Hannah: they confused a generation

Greek Tragedy

Following the success of the Wooden Horse of Troy, the Greeks tried a similar ploy in their attempt to sack the northern city of Corinth, building The Wooden Panda of Corinth. Legend has it that the Corinthians had heard about the wooden horse stunt from a travelling poet, so when the huge panda arrived in the middle of the night, they simply wheeled the beast into the Gulf of Corinth and feasted as it floated away.

The Wooden Panda of Corinth

Picture Credits and Acknowledgements

All images © Getty Images with additional artwork by Mike Ahern, apart from Fact 3 © Corbis, and Fact 75 © DK images.

The Authors wish to thank the following Pandustrators for their generous contribution: Fact 23 illustration by Fergal Brennan; Fact 30 illustration by Chris O'Doherty a.k.a. Reg Mombassa; Fact 49 illustration by Karl Toomey; Fact 64 illustration by Chris Judge; Fact 100 illustration by Mark Wickham; Fact 82 illustration by Grace Chan, Fact 97 illustration by Renate Henschke. The copyright for these illustrations lies with the Authors. Three-dimensional panda-wrangling help, courtesy of Sam Boyd & Enda Loughman.